Plate Tectonics

Our Restless Planet

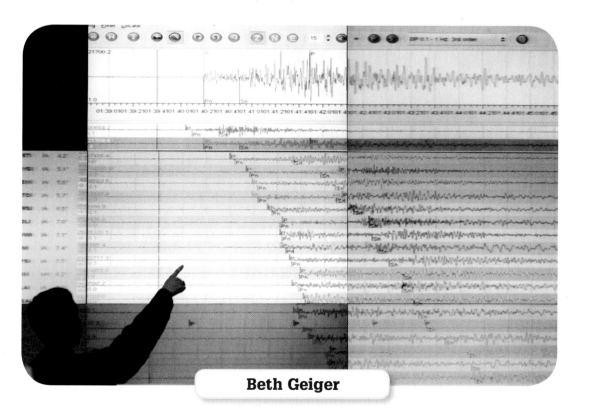

Beth Geiger

Sally Ride, Ph.D., President and Chief Executive Officer;
Tam O'Shaughnessy, Chief Operating Officer and
Executive Vice President; Margaret King, Editor;
Monnee Tong, Design and Picture Editor; Erin Hunter,
Science Illustrator; Brenda Wilson, Editorial Consultant;
Matt McArdle, Editorial Researcher

Program Developer, Kate Boehm Jerome
Program Design, Steve Curtis Design Inc.
www.SCDchicago.com

Sally Ride Science
9191 Towne Centre Drive
Suite L101
San Diego, CA 92122

ISBN: 978-1-933798-46-2

Printed in the United States of America
10 9 8 7 6 5 4 3 2 1
First Edition

Cover: Japan's Mount Fuji is a picture-perfect volcano
that has formed where the Pacific Plate dives beneath the
Eurasian Plate.

Title page: A geologist points to the seismograph
squiggles from a powerful earthquake that rocked Sumatra
Island, Indonesia, in 2009.

Right: The earthquake that struck Sichuan Province,
China, in 2008 ripped apart this road and sent boulders
crashing down on it.

Contents

In Your World

Have you heard the saying, "If you don't like the weather, just wait ten minutes and it will change"?

Weather isn't the only change artist in nature. Plants and animals evolve. Rivers shift course. Even mountains morph.

At least continents stay put, right? Think again! The continent under you is moving. It rides along on one of the plates that make up Earth's **crust**. True, compared to changes in the weather, movement of continents is seriously slow. They creep a few centimeters each year. But add that up over time, and it's plenty. In about 30 million years, Los Angeles will slide past San Francisco. In 50 million years, Africa will bump head-on into Europe.

This slow movement, called **plate tectonics**, is hard to detect. But its effects, like earthquakes and volcanoes, are plain as day. Plate tectonics does more than shake things up a little. Like the clue that helps solve a good mystery, it helps to explain a planet's worth of geology.

POLICE LINE DO NOT CROSS   PO

Plates on the Move

To begin the story of plate tectonics, let's flash back 4.6 billion years. Earth has just formed. It doesn't look much like the cool, blue planet we love. There are no mountains and no trees or grass. There is not even a solid place to stand.

In fact, when Earth formed, it was just a big, hot blob of melted rock. Gradually, like oil separating from water, the lightest stuff rose to the surface. There, it cooled into a layer of solid rock.

Three main layers form present-day Earth. Have you ever eaten a hard-boiled egg? The yolk at the center is surrounded by a soft, white layer and then a hard, thin outer shell. Earth is a little like that egg. At Earth's center is the **core**. The core is extremely hot and made mainly of metals.

Next is the thick **mantle**. The mantle is made of partly melted rock. Finally, there's a cool, thin shell called the crust.

▶ **This map shows how all the plates fit together. Where is your nearest plate boundary?**

▼ **There's a lot more than just dirt beneath your feet. Earth is made up of three major layers.**

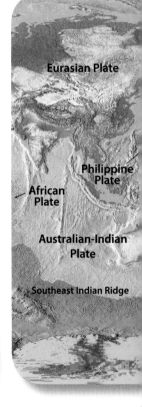

Eurasian Plate

Philippine Plate

African Plate

Australian-Indian Plate

Southeast Indian Ridge

Mantle

Crust

Core

All Cracked Up

Earth's rocky crust is our home, sweet home. What a restless home it is! The crust is broken into a jigsaw puzzle of about a dozen rigid slabs. The slabs are called **tectonic plates**, or often just plates.

Plates are enormous—as big as continents. One, for example, includes almost all of North America. The jigsaw puzzle doesn't stop at the edge of continents, either. Several plates are almost completely covered by oceans, like the one under the Pacific Ocean.

Under the oceans, plates are dense but thin, as little as 5 kilometers (3 miles) thick. Under the continents, the plates are less dense but much thicker—200 kilometers (124 miles) thick in places.

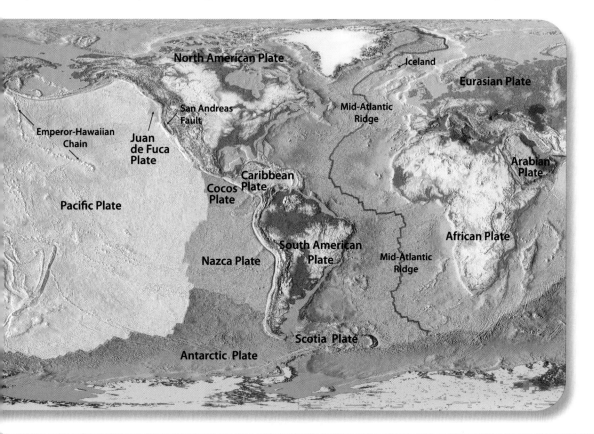

North American Plate
Iceland
Eurasian Plate
San Andreas Fault
Mid-Atlantic Ridge
Emperor-Hawaiian Chain
Juan de Fuca Plate
Arabian Plate
Caribbean Plate
Cocos Plate
Pacific Plate
African Plate
South American Plate
Nazca Plate
Mid-Atlantic Ridge
Scotia Plate
Antarctic Plate

The Bottom Line | Earth's rigid crust is broken into about a dozen huge tectonic plates.

7

Slow but Steady

As Earth's plates move, they bump and jostle each other with incredible force.

Don't reach for a crash helmet, though! Earth's plates move so slowly that they make molasses look fast. The plates creep along anywhere from 1 to 15 centimeters (0.4 to 6 inches) a year. On average, they move about as slowly as your fingernails grow.

Earth's plates don't need to be speedy. They've got time on their side—all the time in the world. The plates have been around at least 1 billion years and maybe much, much longer.

Slow but steady movement adds up. You can prove it by doing a little plate tectonics math. If a plate moves 5 centimeters (2 inches) a year for one million years, how far has it traveled? That's right! It's gone 5 million centimeters (2 million inches)—or 50 kilometers (31 miles)! Some plates have been on the go for 100 million years. In that time, a plate could travel 5,000 kilometers (about 3,100 miles)— more than the distance from New York to Los Angeles.

For a long time, geologists laughed at the idea of continents cruising around like bumper boats. It seemed like a wild idea. Then the evidence started building up.

It doesn't take a scientist to see that the coastlines of South America and Africa fit together like puzzle pieces. This was the first big clue.

▼ **Look at the map on the next page. How did _Mesosaurus_, like the one below, turn up on two continents? No swimming was required. It lived before the two continents split apart 225 million years ago.**

The Case of the Matching Fossils

Fossils provided more clues. Fossils are evidence of past plants and animals and the environments in which they lived. Fossils take many forms. Some are bones, shells, or leaves turned to rock. Others are forms pressed into a rock, like an animal track or leaf shape. All around the world, fossils told strange tales.

Mesosaurus was one such fossil. This little reptile, the size of a cat, lived about 275 million years ago. Fossils of it found in South Africa were identical to those found in Brazil. How could lands so far apart and with such different climates be home to the exact same animals?

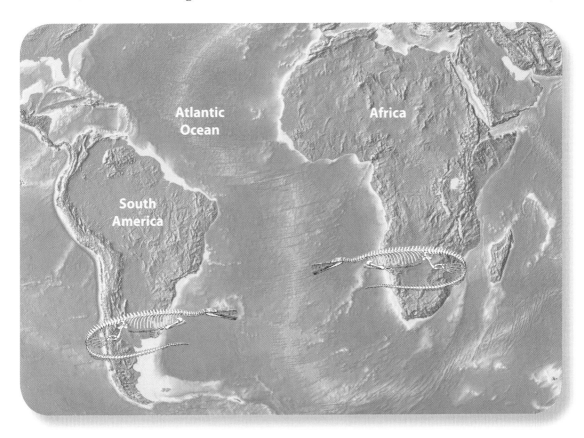

The Bottom Line | Early evidence for plate tectonic motion came from matching continents, fossils, and rock formations.

9

An Ocean That Spreads

Then geologists discovered something even more surprising. They thought oceans were old. So they expected the bottom of the Atlantic Ocean to be blanketed with thick **sediment** built up over hundreds of millions of years.

The geologists were in for a surprise. The sediment was much thinner than they expected. This was especially true along the Mid-Atlantic Ridge—a towering undersea mountain range that runs the whole length of the Atlantic Ocean.

Only one explanation made sense. The crust on the floor of the Atlantic Ocean is young—and still growing. This is called seafloor spreading.

Well, hello there! In 1963, new crust from the Mid-Atlantic Ridge formed the island of Surtsey, off the southern coast of Iceland.

Catching the Action

What could explain all the findings? During the 1960s, the puzzle pieces finally fell into place. Hundreds of millions of years ago, all the continents were lumped together into one giant supercontinent, called **Pangea**.

No wonder the coastlines of South America and Africa fit together so nicely. No wonder fossils from both continents are so similar. These places had once been attached. You could have walked from Sydney, Australia, to London, England, without getting your feet wet. That explains the Mid-Atlantic Ridge. It is a seam where two plates are moving apart.

These days, no one doubts plate tectonics. Earth's plates are on the go, carrying the continents with them as they move.

About 200 million years ago, the Atlantic Ocean was just a little bay between the continents of Africa, Europe, and North and South America. Find where the continents were once attached on the map below. Pangea slowly broke apart, and the bay grew bigger and bigger. That's why today it's an enormous ocean.

200 Million Years Ago

90 Million Years Ago

Present Day

The Bottom Line

The discovery of seafloor spreading in the Atlantic Ocean confirmed the theory of plate tectonics.

Big Forces, Big Effects

What on Earth could be powerful enough to move continents? Geologists believe that **convection** is the driving force.

You can cook up your own example of convection. Watch a pot of thick soup slowly heat on the stove. When soup near the bottom gets hot, it becomes less dense. It rises. Then it cools off and sinks. The circular motion that results is convection. No stirring is necessary.

Of course, Earth's mantle isn't made of potatoes and broth.

But it is partly melted, and it behaves a little like the soup. Heat from deep inside Earth acts as the stove. The heat sets up convection within the mantle. What about the plates? Imagine dropping a crusty piece of toast into the pot of soup. The toast doesn't just float motionless. It moves around as the soup circulates. Now imagine that the toast is a tiny tectonic plate. It's easy to see how convection can push it around.

◄ Earth's massive tectonic plates float on the molten mantle underneath.

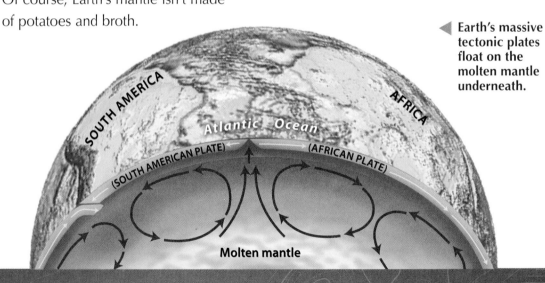

SOUTH AMERICA

AFRICA

Atlantic Ocean

(SOUTH AMERICAN PLATE)

(AFRICAN PLATE)

Molten mantle

Slow-motion Makeover

This steady plate motion gives Earth's surface a slow-motion makeover. The effects are definitely dramatic. They are also super slow. Even the fastest plate moves just 15 centimeters (6 inches) each year. That's less than the width of this book. Most of the action occurs along the edges of plates, where plates meet.

▼ The Nazca Plate moves away from the Pacific Plate at a world-record speed of 15 centimeters (6 inches) per year. Easter Island and its famous statues ride along.

The Bottom Line

Tectonic plates slowly move, floating on the hot, molten mantle. Convection in the mantle drives this motion.

Out with the Old

Speaking of a makeover—remember the Mid-Atlantic Ridge in the Atlantic Ocean? It's a crust-making factory. **Magma** oozes up through the rift—the crack in the crust where plates are splitting apart. The magma then cools into new crust. As this new rock forms, the old rock is pushed to the sides. *Presto*! The plates spread.

This same oozing and spreading happens all along the long undersea ridge that snakes around Earth. Wait a minute. If new crust keeps forming, wouldn't Earth keep getting bigger? It seems like it would, but it doesn't.

As fast as new crust is created, old crust is recycled. In some places where two plates collide, one is pushed down under the other. This mega-recycling process is called **subduction**. It creates deep trenches in the ocean floor. The old crust goes down and gradually melts back into the mantle. And you thought recycling was a new idea! Earth's been doing it for at least a billion years.

Splitsville

Tectonic plates don't exactly steer. They jostle and collide with each other. They rip and split apart. All this banging around shapes and reshapes Earth's mountains and oceans.

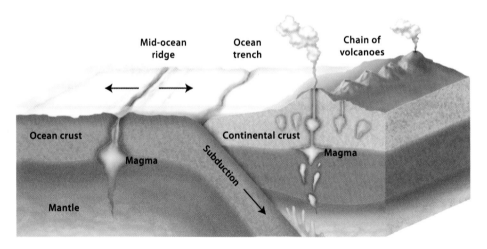

▲ Caution, plates at work! New crust forms and spreads from mid-ocean ridges. Meanwhile, old crust is recycled as it dives under other plates.

What happens when two plates split apart? At the Mid-Atlantic Ridge, magma pushes the North American Plate and the Eurasian Plate away from each other. They are spreading apart about 2.5 centimeters (1 inch) a year.

You can't see most of the Mid-Atlantic Ridge. It's deep underwater. You can, however, see other places where plates spread apart from each other. In East Africa, sizzling-hot volcanoes and sharp rifts mark the line where Africa is pulling apart.

Mid-Atlantic Ridge

Iceland

Mid-Atlantic Ridge

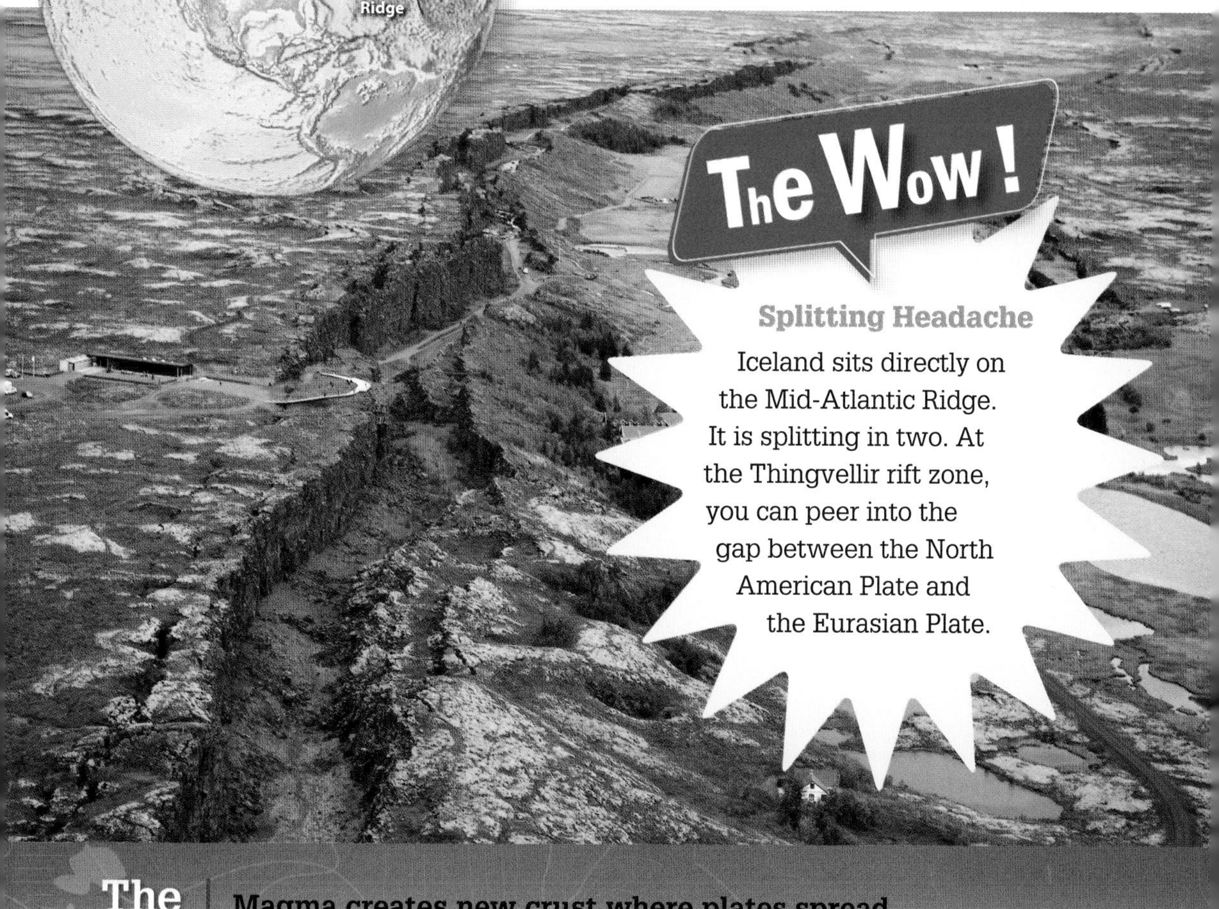

The Wow!

Splitting Headache

Iceland sits directly on the Mid-Atlantic Ridge. It is splitting in two. At the Thingvellir rift zone, you can peer into the gap between the North American Plate and the Eurasian Plate.

The Bottom Line

Magma creates new crust where plates spread apart. At the same time, old crust is recycled where plates collide.

Collision Course

Crash! When two plates collide, you can bet the results are big. Over eons, plate collisions form many of Earth's most spectacular mountain ranges. Even the gentle Appalachian Mountains were once lofty peaks pushed up by a plate collision. Erosion gradually wore them down.

Plates collide in different ways. It depends on the type of plates. In some places, an ocean plate collides with a continental plate. Ocean crust is denser than continental crust. When they meet the ocean plate sinks, and the continental crust rides over it. So long, ocean plate! You're headed down! This is happening off the west coast of South America.

When two ocean plates collide, one forces the other down. What happens when two continental plates collide? Neither plate sinks. They just go head to head. *Crunch!*

In one such mega fender bender, India is crashing into the plate carrying Eurasia. The land crumples and pushes up the towering Himalayas—the tallest mountains on Earth.

▼ **Up they go! The Himalayas are pushed 1 centimeter (0.4 inches) higher each year. Geologists can measure this tiny change using the Global Positioning System (GPS).**

Slip Sliding

Sometimes plates just slide past each other. That's happening right now in California. Most of California is on the North American Plate. The Pacific Plate next door is sliding north along the North American Plate. That's not good news for California.

Guess what's riding north aboard the Pacific Plate? Los Angeles. There goes the neighborhood! Actually, though, the Pacific Plate only moves north about 2 centimeters (1 inch) a year. At that rate, Los Angeles won't meet up with San Francisco for another 29 million years.

Talk about a speed trap! Geologists can track if a plate has moved just one millimeter! The trick is using the Global Positioning System (GPS). GPS pairs satellites orbiting 20,000 kilometers (12,427 miles) above Earth with receivers placed near plate boundaries. The satellites measure the precise location of each receiver over and over as they circle Earth. If the location of the receivers changes, geologists will know how far the plate has moved.

This is how geologists detected that Los Angeles creeps north about 2 centimeters (1 inch) a year. Okay, so Los Angeles won't win any races. But just wait a few million years!

▲ California's San Andreas Fault marks the break between the Pacific and North American plates.

The Bottom Line

Where tectonic plates meet, they may collide, or dive under, or slide past each other.

17

Collecting Data

In the 1950s and 1960s, geologists cruised the Atlantic Ocean and recorded the magnetic direction "frozen" into the rocks on the seafloor. Geologists knew the rocks had formed at different times. So it was no surprise that some rocks pointed north and others pointed south.

When a newly formed rock cools and hardens, tiny magnetic mineral grains in it freeze like little compass needles. The mineral grains all point north, right? Not always! Believe it or not, Earth's magnetic field reverses every so often—sometimes after thousands of years, other times after millions of years. **Magnetic north** flip-flops to the south! Rocks that formed during these periods forever "point" south. This peculiar fact explained why the crust on the Atlantic Ocean floor is younger and still growing. How?

What *did* surprise geologists was discovering that the rocks formed a pattern. The bottom of the ocean was striped like a zebra.

The magnetic stripes run parallel to the Mid-Atlantic Ridge. Some stripes are narrow and others are wide. Even more surprising, the pattern of stripes was the same on each side of the ridge.

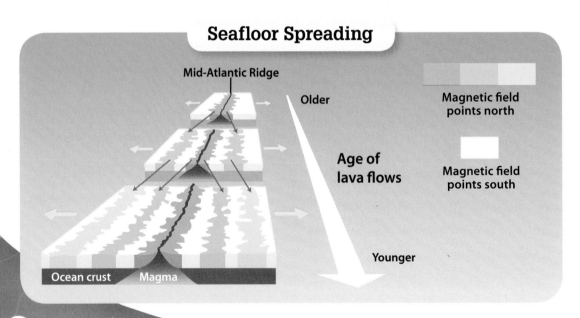

Seafloor Spreading

Mid-Atlantic Ridge

Older

Age of lava flows

Younger

Ocean crust   Magma

Magnetic field points north

Magnetic field points south

Analyzing

What caused the stripes? And why were they identical on both sides of the ridge?

As magma oozes out of the Mid-Atlantic Ridge, it makes new ocean floor. Geologists realized that this new floor was pushing older crust out of the way on both sides of the ridge, making the Atlantic Ocean wider. That meant that the continents were moving apart. *Aha!* The magnetic stripes confirmed plate tectonics. Eventually geologists discovered similar stripes all around Earth, anywhere magma oozes from a mid-ocean ridge.

Your turn! Use the illustration of the Mid-Atlantic Ridge below to answer these questions.

1. How are the magnetic stripes like matching fingerprints?

2. Which color stripe represents the oldest rocks?

3. Which color stripe represents the youngest rocks?

4. Where are the youngest rocks?

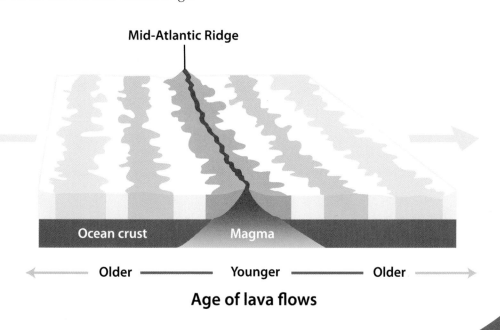

Mid-Atlantic Ridge

Ocean crust · Magma

Older — Younger — Older

Age of lava flows

Magnetic field points north

Magnetic field points south

Action at the Edges

Slow and steady—that's how plates move. But sometimes the effects of this motion are sudden. Not surprisingly, most of the action happens at the plate edges.

When plates move, their motion puts incredible stresses on the crust. It's a little like stretching a rubber band. As you pull and pull, the rubber band stretches and stretches. It stores the energy. Then suddenly—*snap!*—it breaks and the energy is released. *Ouch!*

The crust stores energy in a similar way. When the crust gives, however, the effect is bigger than just a snap.

▼ **In May 2008, an earthquake caused by the Indian Plate's collision with the Eurasian Plate devastated Sichuan Province, China.**

Shake, Rattle, and Roll

Crack! It's an **earthquake**! Earthquakes are sudden shifts in Earth's crust.

Earthquakes occur along fractures called **faults**. When the crust slips, the motion sends energy rippling through the crust.

During an earthquake, the released energy takes the form of **seismic waves**. The seismic waves spread out from the earthquake's **epicenter**. This is the point on Earth's surface directly above where the fault slipped.

How do you measure earthquakes? You use an earthquake scale, of course. The scale, called the **moment magnitude scale**, ranges from 1 to 10. It measures the amount of energy released during an earthquake. But the scale isn't simple. A magnitude-7 quake releases about 32 times more energy than a magnitude-6 quake—and 1,000 times as much as a magnitude-5. You have to be pretty close to the epicenter to feel one measuring less than magnitude 5. If it measures about 6 or greater, then it's rock-and-roll city! The largest earthquake ever recorded was a 9.5, in Chile in 1960.

Earthquakes on the ocean floor can also cause **tsunamis**. These giant ocean waves form when chunks of seafloor buckle up and down during a quake, making the water move up and down, too.

◀ **A view from a helicopter of Phi Phi Don Island, Thailand, in 2004 shows the destruction unleashed by a tsunami.**

The Bottom Line | Plate movements cause stress, or energy, to build up in Earth's crust. The built-up energy is released as an earthquake.

Super Sizzlers

In May 1980, a **volcano** in the state of Washington named Mount St. Helens erupted violently. The blast blew the top off the mountain. Scorching ash shot thousands of meters into the air. Mount St. Helens is one of the Northwest's Cascade Mountains. These beautiful mountains are actually a string of volcanoes.

Here's the inside scoop on these scenic sizzlers. One hundred kilometers (62 miles) below these mountains, the ocean crust is slowly being pushed down. Part of the crust is pushed so deep that it partially melts in the intense heat. The melted rock, or magma, rises. Then, it erupts onto the surface of Earth, where it is called **lava**. The lava cools and hardens into a pile. The pile grows into a mountain. It took just 40,000 years to build Mount St. Helens to a height of 3,000 meters (9,842 feet).

▼ **This is Mount St. Helens in the Cascade Mountains— a string of volcanoes in Washington and Oregon.**

Sometimes, as happened with Mount St. Helens, gas pressure builds up in the magma. Eventually, the volcano explodes like a shaken soda. During the 1980 eruption, Mount St. Helens lost 400 meters (1,312 feet) of height!

Spreading plates build volcanoes, too. As plates spread apart, magma oozes to the surface. The magma cools and piles up. Iceland was formed this way.

Volcano Tour

Volcanoes pop up anywhere magma rises to Earth's surface. Know where one plate is sliding beneath another? Look there and you're likely to find a string of volcanoes. Know where a couple of plates are spreading apart? Look there and you'll probably find another sizzling spot!

You can circle the globe on the hottest tour around. First stop, Japan. Like the Cascades in North America, Japan's line of volcanoes was created by a diving plate. What's different here? Japan isn't on a continent! The volcanoes grow from the ocean floor, forming a string of islands instead.

In the flowery Philippines, Mount Pinatubo erupted big-time in 1991. The culprit? Another diving plate!

Welcome to Mount Etna, Sicily, in southern Italy. Where the African Plate jams north, Mount Etna puts on a lava light show almost every night.

Japanese Islands

▲ Japan is a chain of island volcanoes formed where one plate is diving beneath another.

The Bottom Line | Most of Earth's volcanoes are along the boundaries of plates where there's a lot of jostling.

23

The Ring of Fire

No, the Ring of Fire isn't a circus act. It's way cooler. Look at the map below. The red dots are volcanoes arranged around the Pacific Ocean like a fiery horseshoe. The Ring of Fire includes about 75 percent of the world's active volcanoes. Now that's a real stunt!

Earthquakes jolt the Ring of Fire, too. Scientists once puzzled over this. Now, plate tectonics ties it all together. The volcanoes and earthquakes mark the edges of the Pacific Plate. As the plate moves northwestward, it grinds or dives under other plates along the Pacific Northwest, Alaska, and Japan. The Ring of Fire provides convincing evidence of plate tectonics. It outlines the edge of the Pacific Plate, right where all the action takes place.

The Ring of Fire isn't the only great show on Earth. In fact, volcanoes and earthquakes light up plate edges all around the planet. Just connect the dots!

▼ **The Ring of Fire marks the edges of the Pacific Plate.**

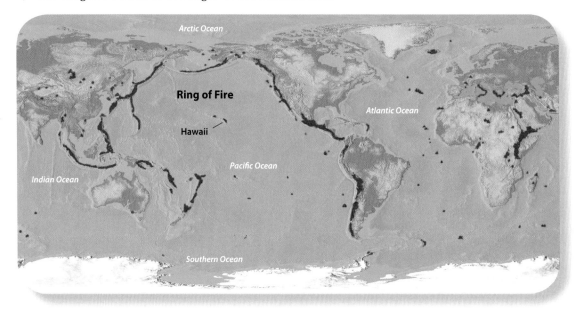

Ready, Set, Go!

Imagine you could drain the oceans. The patterns would become even clearer. You could trace the long mid-ocean ridges, where new crust forms. You could peer into deep trenches where old crust is recycled.

Now, imagine that you could press a button and start Earth's plates moving in slow—*very* slow—motion.

Ready? *Action!* New crust oozes like toothpaste from the mid-ocean ridges. Old crust disappears into deep ocean trenches.

Here come the continents riding along on the plates. Two of them collide. *Crash!* A mountain range pushes up. Other continents pull apart. *Rip!* A new rift forms. Still other continents scrape past each other. *Grind!* The land shifts and shudders with earthquakes. Our restless Earth is nonstop action. And you know why—it's because of plate tectonics.

So THAT'S Why!

What's with the Hawaiian Islands? This volcanic island chain isn't near any plate edges. Instead, Hawaii formed over a hot spot in the mantle under the Pacific Plate. As the plate moves, old volcanoes shift off the hot spot. But the magma keeps oozing. So, a new volcano forms.

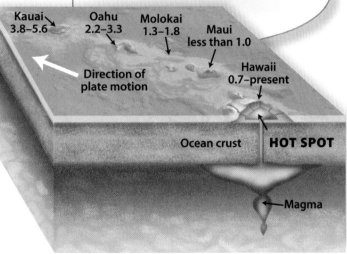

Kauai 3.8–5.6
Oahu 2.2–3.3
Molokai 1.3–1.8
Maui less than 1.0
Hawaii 0.7–present
Direction of plate motion
Ocean crust
HOT SPOT
Magma

▲ The age of each volcanic island is in millions of years. It indicates how long ago each formed over the hot spot.

The Bottom Line

The locations of volcanoes and earthquakes provide strong evidence of plate tectonics.

25

Bracing for the Big One

A scar 1,300 kilometers (800 miles) long splits California. That scar is the San Andreas Fault—a break in Earth's crust where the Pacific and North American plates scrape and grind. As those plates move, stress builds up. When that stored energy is released— *bam!*—an earthquake occurs. Most earthquakes are small—too tiny to feel. But some are huge—big enough to rip trees out of the ground, fling water out of lakes, and knock down buildings.

Earthquakes come in all sizes . . .
Luckily, large earthquakes are rare. In Southern California, the last big one on the San Andreas Fault hit in 1857.

It was a monster—magnitude-7.9—but not a catastrophe. That's because the population at the time was small. Fewer than 40,000 people lived in Southern California. Today, 23 million people call the area home.

The modern-day systems that support all those people have grown, too. Pipes carry water, sewage, and gas to and from homes and businesses. Roads and train tracks crisscross the land. Wires carry electricity and connect telephones and computers. When—and not if—a huge earthquake cuts those lifelines, will people be prepared?

◀ Los Angeles, 1994. When a magnitude-6.7 earthquake hit, utility poles snapped and broken gas lines sent flames through cracked pavement.

Lucy Jones

UNITED STATES GEOLOGICAL SURVEY

Seismologist

◀ Lucy and her sons visit lava fields in Iceland. Volcanic eruptions and earthquakes are part of life on the island nation.

"One of my earliest memories is of an earthquake," says Lucy Jones. She was just two years old at the time. "I remember my mom telling my brother, sister, and me to get into the hallway, curl up, and cover our heads." Earthquakes were just part of life in Southern California. Regular earthquake drills at school sent Lucy and her classmates ducking for cover.

In college, Lucy learned what causes earthquakes—and she was hooked. "I actually read the whole 900-page textbook the first week," Lucy says of geology class. Studying geology taught Lucy how different forces, such as the movement of tectonic plates, shape Earth. Lucy now creates simulations of how future quakes will shake Southern California—and what the consequences will be. That helps other experts prepare. They're strengthening schools, homes and offices. They're stockpiling supplies and teaching kids to take cover—just like little Lucy did during her first quake.

"Earthquakes can strike at any time. So whatever we do to prepare before an earthquake will determine what our lives are like afterwards."

▶ In 1979, Lucy had the opportunity to work in China. Good thing she already spoke fluent Chinese.

Seismologist Lucy Jones has news for the disaster preparation team in her office. A magnitude-7.8 earthquake just hit Southern California. She quickly reports the damage. "The quake sliced freeways and bent railroad tracks. It snapped gas pipelines. It cut phone and Internet connections. Water and sewer lines cracked. The power went out. Buildings collapsed."

Lucy gulps, "Many people died and thousands more were injured. No one answered 911 calls—the phone system was overwhelmed." Lucy pauses, then she reminds everyone it was just an earthquake drill—they were only practicing what they would do in a real emergency. Then suddenly her office shakes for real. A quake? No, just a passing truck.

Seismologists use seismographs to measure the strength and timing of seismic waves. Seismographs are buried around the globe—wherever the Earth shakes. Their measurements zip to a central computer that calculates the earthquake's location and magnitude. The computer automatically posts that information online and then alerts Lucy and other seismologists on their cell phones.

Later, Global Positioning System receivers, like the one below, measure how much the ground moved.

INVENTION CONNECTION

Your Lifelines

Look around your home and neighborhood. Make a drawing of your lifelines. Where do your electricity, gas, and water come from? Where is the nearest cell phone tower? If you're stumped, ask a parent or older sibling for help. Bring your drawing to school. With a partner, brainstorm ways to prepare for an earthquake, hurricane, or tornado.

Quick Quake Changes

In California, engineers and lawmakers learn from earthquakes. California outlawed unreinforced brick buildings after many collapsed in the 1933 Long Beach quake. The 1994 Northridge quake inspired laws because hospitals, like the one below, weren't all quake-resistant. That quake also showed that steel-framed buildings weren't as sturdy as they should be, so engineers designed new ways to connect steel beams.

Hey, I KnoW THAt!

From giant plates and spreading seafloors to sizzling volcanoes and rumbling earthquakes, you've learned a lot about plate tectonics. On a sheet of paper, show what you know as you do the activities and answer the questions.

1. Which city would be best to live in if you really want to avoid earthquakes? Why? (pages 6 and 7)

   a. Boston, Massachusetts
   b. Athens, Greece
   c. Tokyo, Japan
   d. Seattle, Washington

2. Imagine that you are a geologist trying to measure how fast the San Andreas Fault is slipping. Where would you place two GPS receivers? How often would you measure their positions—once a week, once a month, or once a year? Why? (page 17)

3. Which of these mountain ranges is not volcanic? (page 24)

   a. Cascades, North America
   b. Himalayas, Asia
   c. Andes, South America

4. Using this map of the Hawaiian Islands, write down

   a. the direction that the Pacific Plate is moving.
   b. where the next Hawaiian Island will form.
   c. the name of the oldest Hawaiian Island still above water. (page 25)

Formation of Hawaiian Islands

Kauai (3.8–5.6 million years ago)

Oahu (2.2–3.3 million years ago)

Molokai (1.3–1.8 million years ago)

Maui (less than 1.0 million years ago)

Hawaii (0.7 million years ago to present)

Direction of plate motion

Pacific Ocean

N W E S

Glossary

convection **(n.)** the motion of a fluid due to changes in density from changes in temperature (p. 12)

core **(n.)** Earth's innermost layer, located beneath the mantle (p. 6)

crust **(n.)** the relatively thin, rocky outer layer of Earth (p. 5)

earthquake **(n.)** a series of vibrations in Earth's crust produced by the rapid release of energy (p. 21)

epicenter **(n.)** the location on Earth's surface directly above the focus of an earthquake (p. 21)

fault **(n.)** a break in Earth's crust that is caused by the movement of rocks (p. 21)

lava **(n.)** magma that has erupted onto Earth's surface (p. 22)

magma **(n.)** molten rock deep beneath Earth's surface (p. 14)

magnetic north **(n.)** the direction in which a magnetic compass points. It is different from the direction of the North Pole because Earth's magnetic poles do not coincide with its geographical poles. The magnetic pole moves gradually with time in response to Earth's shifting magnetic field. (p. 18)

mantle **(n.)** the hot, partially molten layer of Earth between the core and the crust (p. 6)

moment magnitude scale **(n.)** the scale used to measure the amount of energy released by earthquakes. It is a successor to the Richter scale. (p. 21)

Pangea **(n.)** the supercontinent that existed about 200 million years ago. It broke apart and formed Earth's present continents. (p. 11)

plate tectonics **(n.)** Earth's crust is broken into large sections called plates. Plate tectonics describes their constant motion relative to each other, and explains the geologic activity and recycling of the crust that occur as a result. (p. 5)

sediment **(n.)** loose particles formed by the weathering and erosion of rock (p. 10)

seismic wave **(n.)** a vibration caused by rocks moving and breaking along faults (p. 21)

subduction **(n.)** the thrusting of one tectonic plate under another into the mantle (p. 14)

tectonic plates **(n.)** large sections of Earth's crust that are in constant but very slow motion (p. 7)

tsunami **(n.)** an ocean wave caused by a volcano or landslide or, most often, an earthquake on the bottom of the ocean. When the seafloor shifts up and down during an earthquake, so does the column of water above it. This causes waves that travel through the ocean until they reach land. (p. 21)

volcano **(n.)** a mountain formed from lava (p. 22)

Index

About the Author Beth Geiger, a geologist turned writer, is intrigued by Earth's past and its future. Learn more at www.SallyRideScience.com